What a day! Today is the day Roy moves. Will he make some friends at school?

Roy likes to play with his cat Tiger. They play together in the backyard. Tiger runs over to the soil and digs.

"What can we do today?"
asks Roy.

"Let's go see the stores,"
Mom says.

Mom takes Roy to a toy
store.

Roy looks at many toys in the store. He points to a puppet.

"I like this toy the best," he says. "I can paint this toy!"

Roy's puppet is made from wood. First Roy paints a face on the toy. "I will use oil paint," says Roy.

Mom can fix the puppet's strings. Then she rubs oil on the wood.

"What a great toy!" says Roy.

Roy names his toy Super Boy. "Super Boy can point his arm," Roy says.

Now Roy's puppet is a toy that can move!

Roy takes Super Boy to school. His classmates want to play with Super Boy.

The kids in Roy's class enjoy his toy.

The End

Understanding the Story

Questions are to be read aloud by a teacher or parent.

1. Who has moved to a new place?

2. What kind of store does Roy visit?

3. What does Roy use to paint his puppet's face?

4. What can the puppet's arm do?

5. How do you know that the puppet is fun?

Answers: 1. Roy, his parents, and their cat 2. a toy store 3. oil paint 4. point 5. Possible answer: because Roy's classmates enjoy playing with it

---

Saxon Publishers, Inc.
Editorial: Barbara Place, Julie Webster, Grey Allman, Elisha Mayer
Production: Angela Johnson, Carrie Brown, Cristi Henderson

Brown Publishing Network, Inc.
Editorial: Marie Brown, Gale Clifford, Maryann Dobeck
Art/Design: Trelawney Goodell, Camille Venti, Jillian Gordon
Production: Joseph Hinckley

© Saxon Publishers, Inc., and Lorna Simmons

All rights reserved. No part of this publication may be reproduced, stored in a retrieval system, or transmitted in any form by any means, electronic, mechanical, photocopying, recording, or otherwise, without the prior written permission of the publisher. Address inquiries to Editorial Support Services, Saxon Publishers, Inc., 2600 John Saxon Blvd., Norman, OK 73071.

Printed in the United States of America
ISBN: 1-56577-994-0

Phonetic Concepts Practiced

oi (points)
oy (toy)

ISBN 1-56577-994-0

Grade 1, Decodable Reader 32
First used in Lesson 89

The Storm

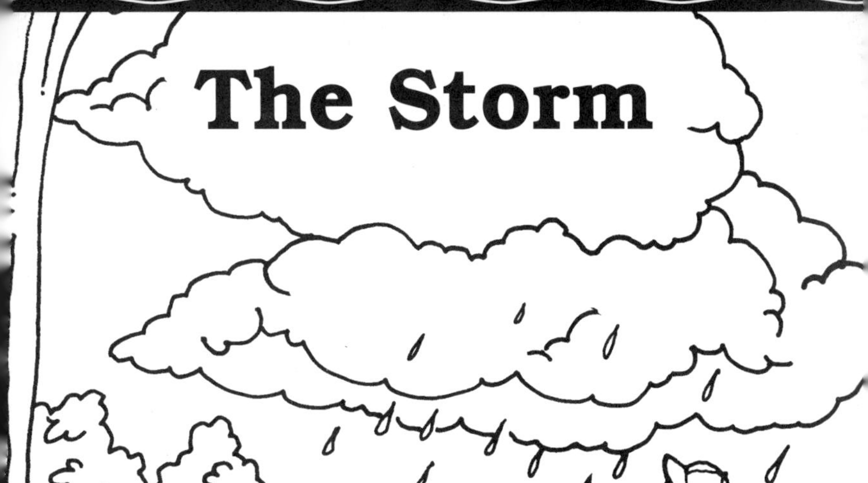

written by Lisa Shulman
illustrated by Janet Skiles

SAXON
PUBLISHERS

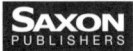

THIS BOOK IS THE PROPERTY OF:

STATE_____	Book No. _____
PROVINCE_____	Enter information
COUNTY_____	in spaces
PARISH_____	to the left as
SCHOOL DISTRICT_____	instructed
OTHER_____	

ISSUED TO	Year Used	CONDITION ISSUED	RETURNED

PUPILS to whom this textbook is issued must not write on any page or mark any part of it in any way, consumable textbooks excepted.

1. Teachers should see that the pupil's name is clearly written in ink in the spaces above in every book issued.
2. The following terms should be used in recording the condition of the book: New; Good; Fair; Poor; Bad.